How To Make Money Using AI

Mwalimu Murithi

Published by Mwalimu Murithi, 2025.

HOW TO MAKE MONEY USING AI

First edition. September 1, 2025.

ISBN: 979-8232187538

Written by Mwalimu Murithi.

How To Earn money with AI

Chapter 1: Introduction to AI and Money-Making Opportunities

Introduction to AI and Money-Making Opportunities

Artificial intelligence (AI) is rapidly transforming the world, and with this transformation comes a plethora of opportunities to generate income. This chapter serves as a foundational introduction to AI, exploring its core concepts and outlining the diverse avenues through which you can leverage its power to earn money. Whether you're a seasoned programmer, a budding entrepreneur, or simply curious about the potential of AI, this chapter will equip you with the knowledge to navigate this exciting landscape.

What is Artificial Intelligence?

At its core, AI refers to the simulation of human intelligence processes by machines, especially computer systems. These processes include learning (acquiring information and rules for using the information), reasoning (using rules to reach approximate or definite conclusions), and self-correction. AI systems are designed to perform tasks that typically require human intelligence, such as visual perception, speech recognition, decision-making, and language translation. These capabilities are achieved through various techniques, including machine learning (ML), deep learning (DL), and natural language processing (NLP).

Key Concepts to Understand

Before diving into money-making opportunities, grasping several fundamental AI concepts is crucial:

Machine Learning (ML): This is a subset of AI where systems learn from data without explicit programming. Instead of relying on pre-defined rules, ML algorithms identify patterns and make predictions based on the input data. This is used extensively in applications like spam filtering, recommendation systems, and fraud detection.

Deep Learning (DL): A more advanced form of ML, DL uses artificial neural networks with multiple layers to analyze data and extract complex features. Deep learning excels in tasks like image recognition, natural language processing, and speech recognition.

Natural Language Processing (NLP): This focuses on enabling computers to understand, interpret, and generate human language. NLP powers chatbots, language translation tools, and sentiment analysis applications.

Data Science: A crucial component of AI, data science involves collecting, cleaning, analyzing, and interpreting data to extract meaningful insights. This is essential for training

AI models and making informed decisions.

Money-Making Opportunities with AI

The applications of AI are vast, creating numerous opportunities to earn money. These opportunities span various skill sets and levels of technical expertise:

AI-Powered Product Development

Developing and selling AI-powered products or services can be highly lucrative. Examples include:

AI-driven mobile apps: Creating apps that leverage AI for features like image recognition, personalized recommendations, or chatbots.

AI-powered software: Developing software solutions for businesses that automate tasks, analyze data, or improve efficiency.

AI-based SaaS (Software as a Service): Offering subscription-based AI-powered software to businesses and individuals.

AI Consulting and Services

As businesses increasingly adopt AI, the demand for AI consultants and service providers is growing. Your expertise in implementing and optimizing AI solutions can be highly valuable.

AI strategy consulting: Advising businesses on how to integrate AI into their operations and achieve their strategic goals.

AI model development and deployment: Building and deploying custom AI models for clients.

AI training and education: Providing training and educational services to individuals and organizations.

AI Freelancing

Platforms like Upwork and Fiverr offer numerous opportunities for freelance AI-related work. You can offer services such as:

Data annotation: Labeling data for training AI models.

AI model development: Creating custom AI models for clients.

AI algorithm optimization: Improving the performance of existing AI models.

NLP tasks: Performing tasks such as text summarization, sentiment analysis, and machine translation.

Investing in AI Stocks

The growth of the AI industry has led to the emergence of many publicly traded companies that are involved in the development and application of AI technologies. Investing in these companies can offer significant returns, but it's crucial to conduct thorough research and understand the risks involved. It is wise to diversify your portfolio to minimize risk.

AI-Driven Content Creation

AI tools are increasingly being used to generate various forms of content, including written articles, images, and videos. While these tools can assist with content creation, human oversight and editing remain crucial to ensure quality and accuracy. You could use these tools to increase your output and make money through platforms like Medium or by offering content creation services to businesses.

Utilizing AI for Personal Finance Management

AI-powered tools can significantly enhance personal finance management. These tools can help automate budgeting, track expenses, and provide personalized financial advice.

Understanding these tools and how to leverage them can lead to better financial management and, ultimately, increased earnings.

Challenges and Considerations

While the opportunities are vast, certain challenges exist:

Technical Expertise: Developing and deploying advanced AI solutions requires significant technical expertise. Continual learning and upskilling are vital.

Data Acquisition and Management: Access to high-quality data is crucial for training effective AI models. Managing and cleaning data can be time-consuming and resource-intensive.

Ethical Considerations: The ethical implications of AI must be carefully considered. Bias in AI models and their potential for misuse are significant concerns.

Competition: The field is becoming increasingly competitive. Standing out requires specialization and a focus on unique value propositions.

Conclusion

AI presents a wealth of exciting opportunities to generate income. By understanding the fundamental concepts, identifying your skills, and continuously adapting to the evolving landscape, you can leverage the power of AI to achieve your financial goals. This chapter serves as a starting point; further exploration and practical experience will be key to your success in this rapidly growing field.

Chapter 2: Lucrative AI Business Ideas

Lucrative AI Business Ideas

The world of artificial intelligence is booming, presenting countless opportunities for entrepreneurs and innovators to generate significant revenue. This chapter delves into a variety of lucrative AI business ideas, ranging from readily accessible ventures to more complex, high-growth opportunities. We'll explore the potential of each, outlining the necessary skills, resources, and market considerations for success.

AI-Powered Content Creation Services

The demand for high-quality content is ever-increasing, across various platforms like websites, blogs, social media, and marketing materials. AI writing tools are rapidly evolving, offering capabilities like generating articles, blog posts, social media captions, and marketing copy. You can leverage these tools to establish a content creation service, catering to businesses, individuals, or marketing agencies. This requires a solid understanding of SEO, content marketing strategies, and the ability to fine-tune AI outputs for quality and accuracy. A strong marketing strategy is crucial for attracting clients and showcasing your service's unique value proposition. To succeed, consider specializing in a niche. Instead of offering general content writing, focus on specific industries like finance, healthcare, or technology.

This allows you to target your marketing efforts more effectively and position yourself as an expert in that area. You can also offer additional services like editing, proofreading, and content optimization to enhance your offerings and command higher prices.

AI-Driven Personalized Learning Platforms

The education sector is ripe for disruption through AI. Develop an AI-powered learning platform that provides personalized learning experiences based on individual student needs and learning styles. This could involve adaptive quizzes, personalized lesson plans, and AI-driven feedback mechanisms. The platform can be targeted at students of all ages and

educational levels, from K-12 to higher education and professional development. This venture requires expertise in AI algorithms, educational pedagogy, and software development. You'll need to build a user-friendly interface, integrate AI models for personalized learning, and ensure the platform's scalability to accommodate a growing user base. Marketing your platform effectively to educational institutions and individual learners is crucial for generating revenue through subscriptions or licensing agreements.

AI-Based Chatbots for Customer Service

Many businesses are incorporating AI-powered chatbots to improve customer service efficiency and reduce operational costs. These chatbots can handle routine inquiries, provide immediate support, and guide customers through various processes. You can offer chatbot development and implementation services to businesses of all sizes, customizing solutions based on their specific needs and industry requirements. Developing effective chatbots requires expertise in natural language processing (NLP), machine learning, and chatbot development platforms. You need to understand conversational design principles to create engaging and user-friendly chatbot interactions. Marketing your services to businesses that value efficient customer support and cost reduction is key to attracting clients.

AI-Powered Data Analysis and Consulting

Businesses are increasingly relying on data-driven decision-making, making AI-powered data analysis a highly lucrative field. You can offer data analysis and consulting services, using AI tools to identify trends, predict outcomes, and provide actionable insights to help businesses make better decisions. This requires strong analytical skills, proficiency in data analysis tools, and a deep understanding of AI algorithms used for predictive modeling and pattern recognition. Specialization within a niche industry can be beneficial. For example, focusing on financial data analysis, healthcare analytics, or marketing analytics can attract a targeted clientele. Networking with businesses and demonstrating your expertise through case studies and testimonials are crucial for gaining credibility and attracting clients.

AI-Enhanced Image and Video Editing Tools

The increasing popularity of visual content across social media and other platforms creates a demand for efficient and high-quality image and video editing tools. Develop or integrate AI-powered features into existing tools to enhance functionalities like image upscaling,

background removal, object detection, video enhancement, and automated editing. You could offer this as a standalone product or integrate it into existing platforms. This venture requires expertise in computer vision, image processing, and software development. Understanding the needs and preferences of your target audience is essential for creating a user-friendly and valuable tool. Marketing your tools through online channels and app stores is vital for reaching potential users and generating revenue through subscriptions, one-time purchases, or in-app purchases.

AI-Powered Cybersecurity Solutions

With the rising threat of cyberattacks, businesses are investing heavily in cybersecurity measures. Develop AI-powered cybersecurity solutions that can detect and prevent threats in real time. This could involve intrusion detection systems, malware detection tools, and security incident response systems. This venture requires deep expertise in cybersecurity, AI algorithms, and software development. This high-growth sector requires staying up-to-date with the latest cybersecurity threats and vulnerabilities. A strong understanding of regulatory compliance and industry best practices is essential. Marketing these specialized services to businesses concerned about data breaches and security vulnerabilities is crucial for success.

Chapter 3: Essential Skills and Resources

Essential Skills and Resources: Your AI-Powered Income Blueprint

The world of artificial intelligence (AI) offers unprecedented opportunities to generate income. However, successfully tapping into this potential requires a strategic blend of skills and resources. This chapter serves as your comprehensive guide, outlining the essential elements you need to navigate this exciting landscape and build a sustainable income stream using AI.

Foundational Technical Skills

While you don't need to be a coding whiz to leverage AI for profit, a foundational understanding of certain technical concepts is invaluable. This section outlines crucial skills, ranging from beginner-friendly to more advanced: Basic Programming: Familiarity with at least one programming language, such as Python (the most popular choice for AI), R, or Java, significantly broadens your opportunities. Many AI tools and platforms rely on coding for customization and automation. Even basic proficiency allows you to tailor pre-built models to your specific needs. Resources like Codecademy, Coursera, and freeCodeCamp offer excellent introductory courses. Data Analysis: Understanding data is paramount in AI. You need skills to collect, clean, and analyze data sets, identify trends, and draw meaningful conclusions. This is crucial for training AI models and interpreting their output. Tools like Excel, SQL, and data visualization software (Tableau, Power BI) are essential. Machine Learning Fundamentals: Gaining a grasp of machine learning (ML) concepts—supervised learning, unsupervised learning, reinforcement learning—enables you to understand how AI models work and make informed decisions about their applications. Online courses from platforms like edX, Udacity, and fast.ai offer comprehensive ML curricula. Cloud Computing: Many AI applications rely on cloud platforms (AWS, Google Cloud, Azure) for their computational power. Familiarity with cloud services allows you to manage and scale your AI projects efficiently and cost-effectively. These platforms usually offer free tiers for beginners.

Specialized Skills for Specific Niches

Beyond the foundational skills, specializing in a particular AI-related area can significantly enhance your earning potential. Here are some lucrative specializations: Natural Language Processing (NLP): If you're interested in text-based applications, NLP is a highly

sought-after skill. It involves building AI systems that understand, interpret, and generate human language. This skill is vital for chatbots, language translation tools, and sentiment analysis systems. Computer Vision: This field focuses on enabling computers to "see" and interpret images and videos. It's crucial for applications like facial recognition, object detection, and medical image analysis. Strong computer vision skills are valuable in autonomous vehicle development and other high-demand sectors. Deep Learning: A subset of machine learning, deep learning utilizes artificial neural networks with multiple layers to analyze complex data patterns. This specialization is essential for tasks like image classification, natural language understanding, and speech recognition. AI Ethics and Governance: As AI becomes increasingly pervasive, understanding its ethical implications and legal ramifications becomes crucial. Experts in AI ethics are needed to ensure responsible development and deployment of AI systems, opening up opportunities in consulting and compliance roles.

Essential Resources: Tools and Platforms

The right resources can significantly accelerate your progress and enhance your productivity. This section highlights crucial tools and platforms: Programming Environments: Integrated Development Environments (IDEs) like Jupyter Notebook, VS Code, and PyCharm provide a user-friendly workspace for writing, debugging, and running your AI code. AI Libraries and Frameworks: Python libraries like TensorFlow, PyTorch, and scikit-learn provide pre-built functions and tools to streamline your AI development. These libraries offer powerful algorithms and functionalities, greatly reducing the development time. Cloud Computing Platforms: AWS SageMaker, Google Cloud AI Platform, and Azure Machine Learning provide cloud-based infrastructure and tools for building, training, and deploying AI models. These platforms offer scalable resources and reduce the need for expensive hardware. Data Sources: Access to high-quality data is critical for training effective AI models. Sources like Kaggle, UCI Machine Learning Repository, and government open data portals offer vast datasets for practice and experimentation. Online Courses and Communities: Numerous online learning platforms (Coursera, edX, Udacity) and online communities (Reddit, Stack Overflow) offer valuable resources for learning, networking, and problem-solving.

Building Your AI-Powered Business

With the right skills and resources, you can leverage AI to build a profitable business. Consider these avenues: AI Consulting: Offer your AI expertise to businesses seeking to integrate AI into their operations. This can involve developing custom AI solutions, providing training, or offering strategic advice. AI-Powered Product Development: Create and sell your own AI-powered products, such as chatbots, image recognition software, or predictive analytics tools. Freelancing: Offer your AI skills on freelancing platforms like Upwork and Fiverr, providing services like data analysis, model building, or AI integration for clients. AI-Driven Content Creation: Use AI tools to assist in content creation, such as writing articles, generating marketing materials, or creating images and videos.

Continuous Learning and Adaptation

The field of AI is constantly evolving. To stay ahead of the curve and maintain your earning potential, continuous learning is essential. Regularly update your skills, explore new technologies, and engage with the AI community to stay informed about the latest trends and opportunities. Attend conferences, participate in online forums, and read industry publications to remain competitive and adapt to the ever-changing AI landscape. This commitment to lifelong learning will be the key to long-term success in the dynamic world of AI-driven income generation.

Chapter 4: Real-World Success Stories

Real-World Success Stories: How Others Are Making Money with AI

This chapter dives deep into real-world examples of how individuals and companies are leveraging AI to generate substantial income. We'll explore diverse applications, from sophisticated algorithms to simpler, readily accessible AI tools, highlighting the potential for financial success across various skill sets and investment levels.

The AI-Powered Freelance Writer

Meet Sarah, a freelance writer who initially struggled to compete with lower-priced writers. Recognizing the growing demand for high-quality content, she invested in learning AI writing tools. She didn't replace her skills; instead, she used AI to enhance them. She uses AI to: Generate initial drafts: AI tools help her quickly create outlines and first drafts, significantly reducing her writing time. This frees her up to focus on editing, refining the tone, and ensuring the content aligns perfectly with the client's brief. Overcome writer's block: When struggling with a specific section, she uses AI to generate alternative phrasing and ideas, sparking her creativity. Improve SEO: Sarah leverages AI tools to analyze keywords and optimize her articles for search engines, leading to increased visibility and more clients. By strategically integrating AI, Sarah increased her output by 40%, allowing her to take on more clients and command higher rates. Her success highlights how AI can empower freelancers to enhance productivity and compete effectively in a challenging market.

The AI-Driven E-commerce Entrepreneur

John, an e-commerce entrepreneur, transformed his business by implementing AI-powered solutions. He initially struggled with customer service and marketing. AI changed everything: Chatbots for 24/7 Customer Support: Implementing AI-powered chatbots significantly reduced response times and allowed him to provide round-the-clock support, improving customer satisfaction and increasing sales. Personalized Marketing Campaigns: He utilizes AI to analyze customer data and create targeted marketing campaigns, leading to higher conversion rates and reduced advertising costs. Inventory Management: AI algorithms predict demand, minimizing stockouts and overstocking, leading to significant cost savings. John's story demonstrates how AI can streamline operations, enhance customer experiences, and ultimately boost profitability in the competitive e-commerce landscape. His willingness to adopt AI technologies has propelled his business to new heights.

The AI-Enhanced Data Analyst

Maria, a data analyst, uses AI tools to enhance her analytical capabilities and unlock greater value from data. She doesn't just analyze data; she uses AI to find insights others miss.

Automated Data Cleaning and Preprocessing: AI tools automate tedious data cleaning tasks, saving Maria significant time and reducing the risk of human error. Advanced Predictive Modeling: She employs AI-powered algorithms to develop predictive models, enabling more accurate forecasting and better-informed business decisions. Data Visualization and Reporting: AI tools help her create compelling data visualizations and reports, effectively communicating insights to stakeholders and driving data-driven decision-making. Maria's success underscores how AI can amplify the skills of data analysts, allowing them to handle larger datasets, identify more nuanced trends, and deliver more valuable insights, leading to increased demand and higher earning potential.

The AI-Powered App Developer

David, an app developer, created a successful mobile application using AI-powered features. He focused on solving a specific problem and found a large, underserved market. AI-Powered Personalization: His app uses AI to personalize the user experience, providing customized recommendations and content, leading to increased user engagement. Machine Learning for Improved Functionality: He incorporated machine learning to continuously improve the app's functionality based on user behavior and feedback. Monetization through In-App Purchases and Ads: The personalized features and improved functionality attracted a large user base, allowing him to monetize the app effectively. David's story highlights how AI can be used to create innovative and engaging apps that meet specific user needs, generating substantial revenue through various monetization strategies. His success is a testament to the power of AI in app development.

The AI-Driven Investment Strategist

In the financial world, sophisticated AI algorithms are being used to analyze market trends and predict future price movements, providing a competitive advantage for investors.

Algorithmic Trading: High-frequency trading firms use AI-powered algorithms to execute trades at lightning speed, capitalizing on fleeting market opportunities. Portfolio Optimization: AI helps investors create diversified portfolios that maximize returns while minimizing risk by analyzing historical data and market patterns. Risk Management: AI algorithms identify and assess potential risks, allowing investors to make better-informed decisions and reduce losses. While entry into this field requires significant expertise, it showcases the lucrative potential of AI in finance. This isn't a 'get-rich-quick' scheme, but careful implementation of AI in investment strategy demonstrates long-term financial success for those with the necessary skills.

Conclusion

These success stories highlight the diverse ways AI can be harnessed to generate income. The key takeaway is that AI is not a replacement for human skills but a powerful tool that augments them, leading to increased productivity, efficiency, and profitability. Whether you're a freelancer, entrepreneur, data analyst, or app developer, exploring AI's potential can unlock new opportunities and pave the way for significant financial success. Remember, continuous learning and adaptation are crucial in this rapidly evolving field.

Chapter 5: Building Your AI Business

Building Your AI Business: From Idea to Income

The allure of artificial intelligence (AI) is undeniable. It promises efficiency, automation, and the potential for groundbreaking innovation. But turning that promise into profit requires a well-defined plan and a deep understanding of the market. This chapter will guide you through the process of building a successful AI-driven business, from conceptualization to launch and beyond.

Phase 1: Idea Generation and Validation

Before diving into coding and algorithms, you need a solid business idea. This isn't about inventing the next groundbreaking AI; it's about identifying a real-world problem that AI can solve effectively and profitably. Consider these avenues: Identify Pain Points: What challenges do businesses or individuals face that could be alleviated with AI? Look at industries you understand – healthcare, finance, marketing, etc. Are there repetitive tasks, inefficient processes, or data-heavy challenges that AI could streamline? Market Research: Once you have an idea, conduct thorough market research. Is there a demand for your proposed solution? Who are your competitors? What are their strengths and weaknesses?

What is the potential market size and revenue projection? Minimum Viable Product (MVP): Develop a basic version of your AI solution. This doesn't need to be perfect; it needs to demonstrate the core functionality and allow you to test your assumptions. The MVP is crucial for gathering user feedback and iterating on your product. Validate Your Idea: Test your MVP with your target audience. Gather feedback, refine your product based on that feedback, and then test it again. This iterative process ensures you're building a product people actually want.

Phase 2: Development and Technical Considerations

With a validated idea, you can move on to the development phase. This involves selecting the right technologies, building your team (if necessary), and managing the development process: Technology Stack: Choosing the appropriate technology stack is paramount. This involves selecting programming languages (Python, R, Java), AI frameworks (TensorFlow, PyTorch, scikit-learn), cloud platforms (AWS, Google Cloud, Azure), and databases. The choice depends on your specific AI solution and your technical expertise. Data Acquisition and Preparation: AI models are only as good as the data they're trained on. Ensure you have access to high-quality, relevant data. Data cleaning, preprocessing, and feature engineering are crucial steps that significantly impact the performance of your AI model. Model Development and Training: This involves choosing the right algorithms and training your AI model using the prepared data. This may require significant computational resources, especially for complex models. Consider using cloud computing services for scalability and cost-effectiveness. Team Building: Depending on your skills and resources, you might need to build a team. This could include data scientists, software engineers, and project managers. Clearly define roles and responsibilities to ensure efficient collaboration.

Phase 3: Deployment and Marketing

Once your AI solution is developed and tested, it's time to deploy and market it: Deployment Strategy: How will your AI solution be delivered to your customers? Will it be a SaaS (Software as a Service) offering, a mobile app, an embedded solution, or something else?

Choose a strategy that aligns with your target audience and business model. Marketing and Sales: Developing a robust marketing strategy is vital. Identify your target audience, craft a compelling message, and utilize appropriate marketing channels (social media, content marketing, paid advertising). Explore different sales strategies, including direct sales, partnerships, and online marketplaces. Customer Support: Provide excellent customer support to build trust and loyalty. Address any issues promptly and efficiently. Regular feedback from customers can also be invaluable for future product development.

Phase 4: Scaling and Monetization

As your AI business grows, you'll need to focus on scaling and monetization: Scalability: Ensure your AI solution can handle increased demand. This might involve optimizing your infrastructure, improving your algorithms, or implementing better resource management strategies. Monetization Strategies: How will you make money? Common strategies include subscription models (SaaS), pay-per-use pricing, licensing fees, and selling AI-powered products or services. Financial Management: Track your income and expenses carefully.

Develop a financial plan to guide your business decisions. Explore funding options such as venture capital, angel investors, or bootstrapping. Continuous Improvement: The AI field is constantly evolving. Stay up-to-date with the latest advancements and adapt your solution accordingly. Continuously collect user feedback and iterate on your product to ensure its long-term success.

Conclusion

Building a successful AI business is a challenging but rewarding endeavor. By following a well-defined process, focusing on a real-world problem, and continuously iterating on your product, you can leverage the power of AI to create a profitable and impactful venture.

Remember that perseverance, adaptability, and a deep understanding of the market are crucial ingredients for success in this exciting and rapidly evolving field.

Chapter 6: Ethical Considerations and Future Outlook

Ethical Considerations and Future Outlook in AI-Driven Earning

The rapid advancement of artificial intelligence (AI) presents unprecedented opportunities for generating income. However, alongside the potential for financial gain lies a crucial responsibility: navigating the ethical considerations that accompany this transformative technology. Ignoring these ethical implications not only undermines the long-term viability of AI-based businesses but also risks causing significant harm to individuals and society as a whole. This chapter explores the key ethical challenges and provides a perspective on the future outlook for those seeking to profit from AI.

Bias and Discrimination

One of the most pressing ethical concerns is the potential for AI systems to perpetuate and even amplify existing societal biases. AI algorithms are trained on data, and if that data reflects biases related to race, gender, religion, or other protected characteristics, the resulting AI system will likely exhibit those same biases in its decision-making. This can lead to discriminatory outcomes in various applications, from loan applications and hiring processes to criminal justice and healthcare. For those building AI-driven businesses, ensuring fairness and mitigating bias requires meticulous data cleaning, careful algorithm design, and ongoing monitoring for discriminatory outcomes. Transparency is crucial; understanding how the AI system arrives at its conclusions allows for the identification and correction of biases.

Privacy and Data Security

Many AI applications rely heavily on the collection and analysis of vast amounts of personal data. This raises serious concerns about privacy and data security. Individuals may be hesitant to share their data if they are unsure how it will be used or if they lack confidence in the security measures in place. Building trust requires implementing robust data protection measures, obtaining informed consent from users, and adhering to data privacy regulations such as GDPR and CCPA. Transparency regarding data usage and security protocols is essential for building a sustainable business model that respects users' privacy rights.

Job Displacement and Economic Inequality

The automation potential of AI has sparked concerns about widespread job displacement. While AI can create new jobs, it may also displace workers in certain sectors. The transition will likely require reskilling and upskilling initiatives to help workers adapt to the changing job market. Those profiting from AI should consider their role in mitigating the potential negative consequences of job displacement, including investing in education and training programs or exploring ways to create a more equitable distribution of the economic benefits derived from AI.

Algorithmic Transparency and Accountability

Many AI systems, particularly deep learning models, are often referred to as "black boxes" due to their complexity and opacity. Understanding how these systems arrive at their decisions can be challenging. Lack of transparency makes it difficult to identify and correct errors or biases, and it can also hinder accountability when things go wrong. Strive for explainable AI (XAI) which aims to make AI decision-making more transparent and understandable. This is not just an ethical imperative but also a practical necessity for building trust and managing risk.

Misinformation and Manipulation

AI can be used to generate realistic-looking but entirely false information, including images, videos, and text. This "deepfake" technology poses significant risks, including the potential for spreading misinformation, manipulating public opinion, and undermining democratic processes. Those creating and deploying AI systems must be mindful of the potential for misuse and should take steps to mitigate the risks of generating or disseminating false information. Developing methods for detecting and combating deepfakes is also crucial.

Intellectual Property and Ownership

The use of AI in creative endeavors, such as generating art, music, or literature, raises questions about intellectual property and ownership. If an AI system creates a work, who owns the copyright? The legal and ethical implications are still evolving, and careful consideration is needed to establish clear guidelines for protecting the intellectual property rights of both creators and AI systems.

Environmental Impact

Training large AI models requires significant computational resources, resulting in substantial energy consumption and carbon emissions. The environmental impact of AI is becoming increasingly important. Those building and deploying AI systems should consider the environmental footprint of their work and explore ways to minimize their carbon emissions, such as using more energy-efficient hardware and algorithms.

Future Outlook

The future of AI-driven earning will depend significantly on how effectively we address the ethical challenges outlined above. A collaborative approach involving policymakers, researchers, developers, and users is crucial to ensure that AI is developed and deployed responsibly. This includes developing ethical guidelines, implementing regulatory frameworks, and fostering a culture of transparency and accountability. Those seeking to profit from AI must prioritize ethical considerations not just as a matter of social responsibility, but as a key component of long-term business success. Businesses that demonstrate a commitment to ethical practices are more likely to build trust with users, attract investors, and maintain a positive public image.

The successful integration of AI into the economy hinges on addressing ethical concerns proactively. By embracing transparency, prioritizing fairness, and focusing on building trust, those who leverage AI can create not only profitable businesses, but also a more equitable and sustainable future for all.

The future will likely see a greater emphasis on explainable AI, robust data privacy measures, and mechanisms for addressing bias and discrimination. Furthermore, increased collaboration between governments, industries, and researchers will be necessary to develop ethical guidelines and regulations that effectively guide the development and application of AI. Ignoring these ethical dimensions risks not only reputational damage and legal repercussions but also the stifling of innovation and the broader adoption of this powerful technology. Ethical AI is not just a moral imperative; it's a strategic advantage in the emerging AI-driven economy.

www.ingramcontent.com/pod-product-compliance
Lightning Source LLC
LaVergne TN
LVHW101618210926
848480LV00009B/823
* 9 7 9 8 2 3 2 1 8 7 5 3 8 *